Implementing the
European CO_2
Commitment

Research by the RIIA's Energy and Environmental Programme is supported by generous contributions of finance and professional advice from the following organizations:

*AEA Technology • Amerada Hess • Arthur D Little
Ashland Oil • British Coal • British Nuclear Fuels
British Petroleum • European Commission
Department of Trade and Industry • Eastern Electricity
Enterprise Oil • ENRON Europe • Exxon • LASMO
Mobil • National Grid • National Power • Nuclear Electric
Overseas Development Administration • PowerGen
Saudi Aramco • Shell • Statoil • St Clements Services
Texaco • Total • Tokyo Electric Power Company*

Implementing the European CO$_2$ Commitment

A Joint Policy Proposal

Helge Ole Bergesen
Michael Grubb
Jean-Charles Hourcade
Jill Jaeger
Alessandro Lanza
Reinhard Loske
Liv Astrid Sverdrup
Angelica Tudini

THE ROYAL INSTITUTE OF
INTERNATIONAL AFFAIRS
Energy and Environmental Programme

First published in Great Britain in 1994 by
Royal Institute of International Affairs, 10 St James's Square, London SW1Y 4LE

Distributed worldwide by
The Brookings Institution, 1775 Massachusetts Avenue NW,
Washington DC 20036-2188

Paperback: ISBN 0 905031 91 1

The Royal Institute of International Affairs is an independent body which promotes the rigorous study of international questions and does not express opinions of its own. The opinions expressed in this publication are the responsibility of the author.

Printed and bound by The Chameleon Press Ltd
Cover by Visible Edge
Illustration by Andy Lovel

Contents

Preface

European policy towards climate change is of global importance. European countries, and the Community, sought to lead the international process that led to the UN Framework Convention on Climate Change, and Europe will host the first meeting of the Conference of Parties to the Convention. The way in which Europe approaches implementation of its collective commitment between diverse member states could set the scene for later global strategies; and could have wider implications for the nature and process of European integration. Yet European climate policy is in serious trouble.

This collaborative report brings together insights from a number of leading European institutes into the dilemmas facing European climate policy. The report argues that Europe does have a clear potential to lead the international process, and could benefit by doing so, but that this will only be achieved by taking a fresh approach at the European level – one that complements and reinforces national efforts. The report proposes an approach that gives a powerful incentive to member states to implement their own commitments, whilst giving them flexibility in how they do so, within a framework that also gives them the option of funding other's efforts when this would be more efficient. It is also an approach that could give European industry the stability, incentive and scope to develop the most innovative and profitable approaches to CO_2 abatement across Europe.

The authors of the report are drawn from five of Europe's leading energy and environmental research institutes. The report expresses the views of the authors, and should not be taken as necessarily indicating consensus across any of the institutes. But we believe the time has come to recognise the need for fresh thinking about climate policy in Europe, and we call upon European policymakers, and other researchers, to consider seriously new options including those set out in this report.

We are grateful to our respective institutes and their supporters, both financial and intellectual, for giving us the opportunity to work upon climate policy and the freedom to develop this report. We are also grateful to those who have helped us to develop the ideas, and in particular to those who attended a review meeting in London and made a range of stimulating comments. It has not been possible to do justice to all their comments, but their varied contributions have helped greatly to deepen the analysis.

November 1994

Helge Ole Bergesen
Michael Grubb
Jean-Charles Hourcade
Jill Jaeger
Alessandro Lanza
Reinhard Loske
Liv Astrid Sverdrup
Angelica Tudini

About the authors and institutions

Michael Grubb is Head of the Energy and Environmental Programme at the Royal Institute of International Affairs, at Chatham House in London. He has published widely on the economic and policy aspects of climate change and related energy policies. The Energy and Environmental Programme is one of seven research programmes at the RIIA, which is an independent, self-governing body that has existed since 1920 to further the study and understanding of all aspects of international affairs.

Helge Ole Bergesen is Senior Research Fellow at the Fridtjof Nansen Institute, Norway, is editor of the *Green Globe Yearbook*, and is Associate Professor at the Norwegian School of Management. Liv Astrid Sverdrup is a researcher at FNI's Energy and Environment Programme specialising in EC studies and climate policy. The Fridtjof Nansen Institute is an independent foundation with a long record of applied research in international relations centred on resource, energy and environment issues.

Jill Jaeger is presently on leave from her position as Programme Leader and Reinhard Loske is senior researcher in the Climate Policy Division at the Wuppertal Institute for Climate, Environment and Energy. The institute was founded in 1991 and concentrates on policy analysis with respect to the issues of climate change, energy systems, material flows and structural change and transport. Dr. Jaeger was recently appointed Acting Deputy Director for Programs at the International Institute for Applied Systems Analysis in Austria.

Alessandro Lanza is Research Director at the Fondazione ENI Enrico Mattei (FEEM), Milan. Angelica Tudini has been a researcher at FEEM since 1991, and is a member of the Italian Ministry of Environment's Expert Group on international activities. The Fondazione has since its establishment in 1990 become Italy's leading institute promoting research and policy analysis in the field of energy, environment and economic development.

Jean-Charles Hourcade is Director of the Centre International de Recherche sur l'Environment et Développement, France. CIRED is both part of the Ecole des Hautes Etudes en Sciences Sociales (EHESS – Paris) and of the Centre National de la Recherche Scientifique (CNRS – France). Hourcade spearheaded Oïkia – an official network of 12 French research teams, mainly within social sciences, to work cooperatively on climate change.

Executive summary

European environmental policy faces its biggest test. The Maastricht Treaty established a high level of environmental protection as one of the central aims of the Union. Scientific evidence is accumulating that growing CO_2 emissions may threaten the stability of the world's climatic systems. The European Council of Ministers has declared and repeatedly reaffirmed that CO_2 emissions from the European Union in the year 2000 should be stabilised at 1990 levels, and the Union itself and most member states have ratified the UN Framework Convention on Climate Change on that or related bases.

The European position is critically important internationally. Failure to achieve the goal could have serious repercussions for global efforts to contain the problem, particularly concerning the involvement of developing countries. Conversely, successful implementation of a coherent strategy between the diverse members of the EU could have a powerful demonstration effect. The political leadership, and the economic benefits which could flow from sharing efforts and markets for technical innovation and improved energy efficiency, could establish a basis for an enlarging international coalition with mutual benefit.

But the present strategy is failing. The main initiatives proposed by the European Commission for European-wide climate/energy policy measures have been rejected or weakened, on the grounds of subsidiarity and competitiveness, to the point where they can make little real contribution to the declared objective or to longer term reductions. The national emission targets declared by member states should, if met, bring the Union close to its collective goal; but the first report under the Monitoring Decision, which was established to review whether national actions are sufficient to achieve the national and European goals, suggests that they are not. Expansion of the Union further increases the difficulty of achieving the objective and highlights questions about efficient distribution of the effort. Current policies seem even less likely to achieve the emission reductions that are likely to be required beyond the year 2000.

The central challenge is that policies need to be implemented at many different levels. The key energy policy decisions required to meet European-level CO_2

commitments cannot all be taken centrally, but an incentive is required to ensure that member states take their contribution to the European total seriously, whilst recognising national differences in energy structures and policies. This requires a framework that penalises failure to meet agreed national targets and rewards over-achievement.

This can be achieved if the member states agree (a) to share the European target into national ones, which probably requires only minor modification of existing national targets; and (b) to establish a legal framework that allows member states to exchange these initial targets on mutually agreed terms, so that one country can buy additional quotas from another that proves more easily able to reach, and with the added incentive exceed, its initial target. Allowing such exchanges will greatly enhance the effectiveness, efficiency and feasibility of this approach as compared with one that does not provide any framework for handling variations from the intially agreed targets.

Such a system would be fully compatable with other policy measures at both national and European levels, and accords with established principles of Subsidiarity and Polluter Pays. By creating an efficient framework of incentives, it would indeed encourage other efficient control measures to be considered more seriously. Establishing such a European-wide market-based system for CO_2 constraint could ultimately reduce the regulatory burden on European industries and stimulate more innovative and ultimately profitable responses by them.

A number of options for the detailed design of such a system exist. The fact that emission constraints are likely to be extended beyond the year 2000 increases the value of such a system, and improves its characteristics. The scope can readily be extended to allow for expansion to other interested parties with mutual benefit. Many aspects of design are a matter of choice to be determined by analysis and negotiation among member states, but the overall feasibility of such a system depends primarily upon political will.

Negotiations should be opened to develop such a system as part of the EU strategy for implementing its emission goals. A commitment to this, in the run-up to the first meeting of Parties to the UN Climate Convention, would bring many benefits, and would allay growing suspicions that the CO_2 declarations by European countries and institutions have been symbols without substance.

Chapter 1

Introduction: the basis for climate policy in Europe

The threat of climate change poses some of the most important and politically difficult issues for energy and environmental policy in the European Union. Commitments have been made, both within the Union and globally under UN auspices. Currently, the will and means to implement these commitments are in doubt. This report, prepared between researchers at five of Europe's leading policy research institutes, summarises the current situation and proposes ways forward.

The underlying issues are well known. On the one hand there are uncertain risks to the global climate system, on which ultimately we all depend, from rising concentrations of greenhouse gases such as carbon dioxide (CO_2). On the other hand there are the processes which emit greenhouse gases, notably energy from fossil fuels, on which we also depend. Policy needs to strike a sensible balance, with recognition that the central requirement is for consistent and efficient implementation of a long-term strategy that meets existing commitments, ensures the sustainable evolution of European energy systems, and is well founded on a good understanding of both the economic and scientific issues.

1.1 Scientific basis

At one level the science is straightforward. The fact that greenhouse gases in the atmosphere trap heat near the surface is both common sense, and one of the best verified facts in climate science. Nor is it disputed that human activities are increasing the concentration of greenhouse gases, though the exact extent of human contributions and the lifetime of the different gases in the atmosphere is still debated. The biggest uncertainties arise from the fact that heating the earth's surface does a lot of other things as well. Changes in water vapour, clouds and ice cover, can either reinforce or offset the original impact. Warming, together with rising CO_2 levels and ocean levels, is also likely to affect plant species and productivity, and atmospheric and oceanic circulation patterns, so that regional changes may be very different from the global average.

Thus, climate change itself is not an uncertain hypothesis but a virtual certainty: what we don't understand in detail are its likely degree, rate, consequences, and timescales. In 1990 the first report of the Intergovernmental Panel on Climate Change (IPCC), established by governments to examine the issue amid competing claims, demonstrated a widespread scientific consensus on the nature of the problem and the associated uncertainties. Scientific developments since the first IPCC report have not substantially altered the scientific basis for policy, though fuller analysis of the role of non-CO_2 trace gases like sulphur dioxide (SO_2) and CFCs helped to lower the central projection of the rate of average global temperature change by 20-30% in the 1992 supplement to the 1990 report. A second update on radiative forcing in 1994 reinforces this estimate.

The IPCC projected the likely global average temperature increase over the next century at the upper extreme to be almost as large as the change from the last ice age until the present (4.5°C), and at the lower end of the range to be about a third of that (1.5°C), with best estimate about 2.5°C. The regional patterns of change are still more uncertain, though most models predict a net drying in continental interiors, and changes in the nature and distribution of extreme events. It is for example possible (though far from certain) that past greenhouse gas emissions have contributed to the growing severity of droughts in Africa and elsewhere.

Conclusive observational proof that humanity is changing the climate (i.e. beyond the conceivable range of natural variation) is not expected for some years. But model simulations and actual observation over the past few decades appear increasingly consistent, increasing confidence in our basic scientific understanding and projections of global average changes.[1]

---

[1] The potential impact of sulphur emissions in offsetting greenhouse warming, and to an extent improved understanding of the role of CFCs, helps to explain the lack of global average temperature rise over the period 1940-1970. Increased knowledge about the spatial and temporal distribution of temperature change is consistent with this understanding. There remain many detailed aspects of the temperature record (including differences between ground based and satellite measurements) that are not fully explained; the complexity of the system and number of different influences is such that detailed regional and year-to-year fluctuations may never be fully understood. Trends in the last couple of years have been influenced by the collapse of emissions from Eastern Europe, the Mount Pinatubo eruption, and the El Niño fluctuation in Pacific ocean currents.

Analysis of temperature data over the past hundred thousand years has both confirmed that the recent atmospheric changes are unprecedented, and revealed that in some circumstances the global climate system has been unstable, with rapid shifts of climatic patterns probably corresponding to changes in oceanic circulation. Scientists are currently debating data suggesting that climatic instabilities may have occurred in the warm period before the last ice age (when the average temperature was about 2°C warmer than today); and other data which appear to indicate a substantial decline in one of the four main ocean 'sinks' in recent years.[2] Scientific developments since 1990 thus maintain, and perhaps increase, the basis for concern and precautionary action.

1.2 Economic and strategic basis

The costs of climate change itself are highly uncertain. If climate change is smooth and reasonably predictable for given regions, the direct economic costs of adapting to it may be modest, especially for developed countries in temperate regions. The extent to which natural ecosystems could adapt is still a matter of debate, and there could be other important forms of non-economic costs – such as increased pressures for migration from countries which are hotter, very low lying, or otherwise more severely impacted. But few scientists consider it reasonable to assume smooth and predictable change. The nature of the global system makes it hard to predict changes at the regional level, so the scope for preparing against future climatic change is limited, particularly if much of the damage is related to extreme events such as storms or recurrent droughts. In addition, many scientists consider that changes may be marked

---

[2] Rather than detail here the large number of scientific papers, and related commentaries, that underpin these developments, the reader is referred to the work of the Intergovernmental Panel on Climate Change. The First Scientific Assessment was published as: J.T.Houghton, G.J.Jenkins and J.J.Ephraums, eds, *Climate Change: the IPCC scientific assessment*, Cambridge University Press, 1990. The 1992 update was published as J.T.Houghton, B.A.Callander and S.K.Varney, eds, *Climate Change 1992*, CUP, 1992. The 1994 Special Report on Radiative Forcing is also due to be published by CUP. Draft chapters for the full Second Assessment Report by the IPCC are currently being circulated for peer review, and the final report in 1995 will provide a comprehensive reassessment of the climate issue. The source papers and commentaries on most scientific developments can be found in the journals *Nature* (UK) and *Science* (US).

by relatively rapid changes from one climatic state to another. This could occur at the regional level, and even at the global level if the major patterns of ocean current circulation should shift.

The real policy dilemma is that by the time we can be certain about such changes, even if we then reduce emissions quite rapidly we will still be committed to many further decades of change before the problem can be brought under control, because of the immense inertia in the global system. Thus we have to act whilst the extent and implications of the problem are still uncertain, as recognised in the 'precautionary principle' accepted in the UN Climate Convention. The Convention declared the objective to be:

> stabilization of greenhouse gas concentrations in the atmosphere at a level that would prevent dangerous anthropogenic interference with the climate system within a time frame sufficient to allow ecosystems to adapt naturally to climate change, to ensure that food production is not threatened, and to enable economic development to proceed in a sustainable manner.

The appropriate level and rate are, for the reasons summarised above, highly uncertain, but the 1994 update of the IPCC report concludes that stabilizing CO_2 concentrations at levels anywhere below about three times pre-industrial concentrations will probably require global emissions to be reduced substantially below current levels.

There is much debate about the potential costs of limiting emissions, especially of CO_2. A certain amount can be done very cheaply (or even at negative costs), primarily by exploiting the cost-effective potential for improvements in energy efficiency, but also by using more natural gas to displace coal, and special applications of renewable energy sources where they are already cost-effective. Steady reforms of subsidy and tax systems can play an important role in helping to induce such changes, and to accelerate technical advances. Evidence and experience suggests that there is considerable potential for technical improvements in new energy technologies, if there are sufficient incentives for industry to invest in their development and expanded application over a sustained period; this would also make the industries well placed competitively if and when more countries have to adopt greenhouse gas constraints, as discussed in Chapter 2.

Delaying action incurs risks, both environmental and industrial: environmental risks, because the risks are likely to increase with the maximum rate and extent of atmospheric change; industrial risks, because of the possibility that we will then be forced to curtail emissions rapidly to avoid more serious planetary consequences.

1.3 The European position and the UN Climate Convention

The European Community took a forward stance on climate change since the problem was first formally recognised as a serious international policy issue in the late 1980s. The decision with the single greatest impact was the statement by the joint Council of EC Energy and Environment Ministers of Member States, on 29 October 1990, that the Community as a whole would stabilise its total CO_2 emissions at 1990 levels by the year 2000 (see Chapter 3).

In the weeks after this declaration, the UN agreed to launch negotiations on an international convention. The resulting UN Framework Convention on Climate Change commits industrialsed country Parties to adopt policies 'with the aim of returning individually or jointly to their 1990 levels' of emissions of CO_2 and other greenhouse gases, and to communicate to the Conference of Parties established by the Convention detailed information on the measures being adopted for achieving this goal.[3]

The Convention was signed at the Rio 'Earth Summit' by 152 countries, including all EC members, plus the European Community itself.[4] Most member states have ratified the treaty individually and, as the European Community,

---

[3] The actual wording in the Convention is convoluted and deliberately ambiguous in key parts, owing to the refusal of the Bush Administration to accept clear legal obligations on emission targets in the treaty (a summary is given in M.Grubb et al., *The Earth Summit Agreements: a guide and assessment*, RIIA/Earthscan, London; Brookings, Washington, 1993). In public declarations, however, it has been accepted by most participants – subsequently including the USA – as a commitment to ensure that CO_2 emissions in the year 2000 are not higher than in 1990.

[4] In this discussion, we use the term Industrialised Country Party to include the European Community as a Party to the Climate Convention in its own right. The Community remains the relevant legal entity after the Maastricht Treaty in 1993 established the European Union as the political entity of states that are members of the European Community. In this report we refer to the European Community (EC) and its specific commitments and policies in this context; use of the term European Union is reserved for more general references.

did so jointly in December 1993. The Convention entered force in March 1994. The USA has also ratified the Convention and declared likewise that it will return its emissions of greenhouse gases to 1990 levels by the year 2000. The European Council has consistently reaffirmed its commitment to this goal, and most developing countries now regard such action by industrialised countries as a precondition for any consideration of substantive action or strengthening of the Convention on their part.

All these processes have thus lent great political and legal weight to the objective of stabilising EC CO_2 emissions. To renege upon it now would undermine the UN Convention in the very area in which the EC fought hardest for stronger wording, and would be used by developing countries as a *prima facie* reason why they should not take significant action. Given that the target has also been repeatedly reaffirmed by the Council of Ministers, it would also make the EC look foolish and lacking in seriousness about such undertakings.

Furthermore, Europe is in many ways a microcosm of the global problem. There is a North-South element, with recognition that growing emissions from the less developed members of the Union will have to be accommodated by reductions in some of the richest members. There is the tension between sovereignty, the needs of collective responsibility, and the harmonisation of responses where appropriate according to the principle of subsidiarity. Yet the European Community is a uniquely well developed international structure for the resolution of such issues in the pursuit of common goals. If the EC cannot effectively implement its undertaking to stabilise CO_2 emissions, it is unlikely that any effective solution can be wrought globally. Conversely, an effective EC strategy could be of immense benefit as an international demonstration of ways forward.

Yet the EC does not at present have a convincing strategy for achieving its commitments. In 1990, the Council of Ministers asked the European Commission to prepare a Community-wide strategy to stabilize EC CO_2 emissions. In the resulting proposals,[5] a key role was given to a programme

---

[5] 'A community strategy to limit carbon dioxide emissions and to improve energy efficiency', COM(92) 246 final, Brussels 1 June 1992. A detailed economic appraisal of the Community's strategy, spanning both international dimensions and the specific instruments proposed, is given in *European Economy* no. 51, CEC-DGII (Economic and Financial Affairs), Brussels, May 1992.

of EC energy efficiency standards ('SAVE'), and proposals for a harmonised carbon/energy tax, together with other measures outlined in Chapter 3. For reasons analyzed more fully in Chapter 3, it is now clear that these measures will now not be implemented in anything like the originally proposed form or timescale. Nor (as also discussed in Chapter 3) does it seem likely that national measures currently being taken will achieve EC CO_2 stabilisation, let alone longer-term emission reductions. Four years after the resounding declarations that accompanied the stabilisation declaration by the EC, and similar declarations by many member states, policies to harmonise or even coordinate responses, or to achieve the collective EC goal, seem remote.

The Climate Convention requires that all industrialised country Parties to the Convention should submit reports to the Secretariat, detailing the 'policies and programmes' being implemented to achieve the stabilisation goal, by 21 September 1994. The European Community, and some of its member states, failed to do so; an outline report from the Commission is now due to be submitted by the end of the year but it will not contian convincing detail on progress towards stabilising EC CO_2 emissions. It is an embarrassment to the EC and its claim to lead on the climate issue, and to the member states who gave the EC authority to sign. It is particularly awkward for the less developed members, who signed on to the Convention's stabilisation goal explicitly under the EC's umbrella target. And it bodes ill for the prospects of forging an efficient longer-term European response – let alone one at the global level.

In this short report we examine the status of EC climate policy and propose a way out of the policy dilemma. It is not a proposal which provides a magic formula for avoiding difficult decisions, or which can resolve all debates on other aspects of climate policy. But it is, we contend, an approach that could break the stalemate in Europe and stimulate a range of beneficial developments for Europe, and perhaps even provide a model which could form the nucleus of an effective global approach to limiting the threat of climate change.

Chapter 2 of this report puts the European situation with respect to CO_2 in the global context. Chapter 3 then examines the development of EC climate policy in Europe to date, and the reasons for the current impasse. Chapter 4 considers the impact that the imminent enlargement of the Union may have on climate policy. In Chapter 5, the basic proposal is outlined, and Chapter 6 then considers some of the more detailed issues involved in developing such an approach.

Chapter 2

European climate policy in a global context

Climate change is a global problem. Ultimately, action to arrest global warming must involve all the major regions of the world, and all the major sources of greenhouse gases. At present however, the world is in the early stages of addressing the problem, and global solutions cannot spontaneously arise; they must be constructed upon the efforts and leadership of key participants, with an eye to the international repercussions of their actions – or failure to act.

This chapter addresses the role of the EC in this context, focusing upon CO_2 emissions from fossil fuels. This focus arises because CO_2 emissions form the most important single contribution to the climate problem, and the great majority of EC emissions arise from fossil fuel combustion. Other sources, and other gases, make significant global contributions, and will need to be curtailed. But the climate problem cannot be addressed meaningfully without tackling fossil fuel emissions, which also raises some of the most difficult and important political and economic issues. Hence the central place accorded to fossil fuel CO_2 emissions in the overall climate debate.

2.1 European CO_2 emissions and the global context

The collapse of the Soviet Union has left the European Union as the second biggest cohesive economic group in the world, and the second biggest emitter of CO_2 after the United States. In 1993 the USA accounted for 25% of global CO_2 emissions from fossil fuel, and the EU-12 (the twelve members of the EU in 1994) for 14.5%; the accession of Austria and Scandinavian countries in 1995 will bring this above 16%. The countries of Central Europe and the former Soviet Union together accounted for another 17% of fossil fuel CO_2 emissions, about half of these being from Russia itself. Emissions from developing countries are rising rapidly, and now account for over a quarter of the global total. If this emissions growth continues unabated, within 30-40 years the developing countries could account for more than half the global total fossil CO_2 emissions.

Emissions per capita are compared in Figure 2.1, which shows the main countries individually and groups together others with similar per capita emission levels. Among the main industrial groups, the order remains similar: the USA is most profligate by this measure, followed by Canada and Australia; per capita emission from the former USSR had by 1993 fallen towards European levels; and the EU average is similar to that of Japan. There is, however, a big gulf compared with most developing countries, where fossil CO_2 emissions per capita are typically several times lower than in developed countries.

Because of this gulf, and the fact that the developed countries have dominated emissions historically, developing countries have taken the attitude that stabilisation of emissions from industrialised countries is a precondition for them to consider any substantive abatement action. The economic collapse of the former Soviet Union means that emissions are contracting there anyway, and precludes them from taking a more active position. The focus is thus upon OECD countries.

The USA has already published its national strategy for 1990-2000 stabilisation[6] of greenhouse gases. Japan has committed to a target of per capita 1990-2000 CO_2 stabilisation, and published an Action Plan to Arrest Global Warming, and is working on more detailed measures. Japan has also emphasised the long-term nature of the problem, and technological strategies towards this long term. There is thus a strong spotlight upon Europe, which led the declarations of CO_2 emissions stabilisation but which has yet to agree any coherent strategy for achieving it.

European action is particularly important because of the international history outlined in Chapter 1, and for a range of other reasons. One is simply the global importance of the European energy industries. The EU does, or with enlargement will, host many energy-related industries of global importance:

---

[6] We use this term to describe the aim that emissions in the year 2000 should return to their 1990 levels. A number of national targets (eg. USA, UK) are of this character and the present Convention "aim" is interpreted in this sense. Stabilisation in full implies a commitment that emissions should not rise again thereafter. The EC declaration refers explicitly to "stabilisation" (see Chapter 3).

Figure 2.1 CO$_2$ emissions per capita, 1993

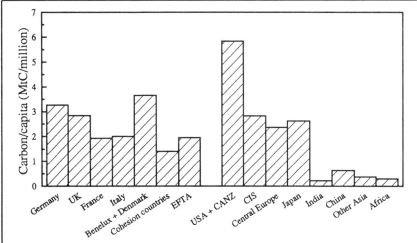

Note: The figure shows fossil CO$_2$ emissions per capita (as tonnes of carbon per person in 1993) for major countries and groups:
Benelux = Belgium, Netherlands, Luxembourg;
Cohesion countries = Greece, Spain, Portugal, Ireland;
CANZ = Canada, Australia, New Zealand;
EFTA = Austria, Finland, Norway, Sweden, Switzerland.
Source: derived from BP Statistical Review of World Energy, 1994, using emission factors from M.Grubb et al, *Energy Policies and the Greenhouse Effect: Volume II*, RIIA, London (Appendix II).

international oil and gas companies such as Shell, BP, AGIP and Elf, and gas companies such as British Gas, Gaz de France, Ruhrgas and SNAM that are becoming increasingly international; equipment manufacturers such as Siemens, ABB and Babcock; and increasingly internationalised electricity companies such as the major UK generators and grid companies, as well as a wide variety of manufacturers of equipment which use energy, from vehicle to appliance companies. The perceptions, technological capabilities and attitudes of these companies will be affected by how seriously the EC seeks to implement its CO$_2$ commitment, and this in turn will affect the kind of technologies and processes they export – and will be able to export if and as other countries take actions to limit emissions.

2.2 The EU as a microcosm of the global issues

Another reason why the position of the EU and its member countries is important is because in several ways the EU represents a microcosm of the global problem. EU member countries vary with respect to economic and institutional factors, and some of the problems faced by the EU reflect those that could arise, on a larger scale, at the global level in the negotiation of a coordinated climate change strategy.

Economic issues

The economic situation of EU member states resembles, though at a smaller scale, the classical North-South scheme. There are four countries in a less advanced development stage: Spain, Portugal, Ireland and Greece. Their economic situation is the basis for their common position on some aspects of climate change policy: these 'cohesion countries' do not want to bear the responsibility for past emissions of other EU countries, and they fear any constraint on energy consumption as an obstacle to the main aim of economic growth. Moreover, the costs of limiting emissions to 1990 levels would be higher for the currently less developed countries as their economies are likely to grow faster, and start from a lower basis. Climate policy declarations in the EC have recognised the disparity, and have accepted that emissions from these countries are likely to grow in the context of overall EC emissions stabilisation, requiring reductions from some other member states.

Besides the decision on the extent of action, economic factors also affect the choice of policy instruments. The same strategy will have different costs for different countries depending on the economic structure, existing taxation (average fossil fuel prices in EU countries vary mainly because of differing taxes) and resource availability. For example, in the discussion on the carbon/energy tax, the size of the nuclear contribution in France has motivated French support for a tax entirely based on the carbon content of fuels rather on a mix of carbon and energy; the opposite is true for Germany. Also, similar instruments would not be equally easy or efficient in all countries.

Such differences are reflected further in Chapter 3; here, the point is that such differences are mirrored and amplified at the global level, so that attempts to distribute and coordinate action in the EU towards the collective goal is likely to provide valuable lessons for, and may even help to chart a path for, a structure for broader international action.

Institutional factors and subsidiarity

Beside the different economic and technical aspects that divide countries, political and institutional factors further impede the process of policy coordination. Aspects of this include the general national approach to policymaking; specific elements that influence the position on the particular climate change issue; and the state of the debate on sovereignty, subsidiarity and the strength of EC decision-making powers.

European countries vary in the assumptions they bring to policymaking. For example, Wynne[7] draws a distinction between a 'top-down' approach to environmental policy centred on formal policy institutions (e.g. The Netherlands, Germany), and a 'bottom-up' approach (e.g. UK) in which diverse actors other than the formal institutions play an important role in policy development and implementation, rather than simply adapting to it. A wide range of other and more diverse institutional and cultural differences affect policy. Acceptance of policies can also be influenced by other developments; for example, the UK population is most unlikely to accept additional energy taxation just after the government has fought to introduce VAT on domestic energy consumption.

The current EC legal system allows a balance of decision-making through the instruments of Directives that define agreed broad policies while leaving to member states the decision on how to implement them. Therefore the relative weights of centralised and decentralised decisions are defined on a case by case basis. Previous experience – the Large Combustion Plant Directive on SO_2 and the EC common position within the Montreal Protocol – suggests that it is much easier to implement decisions on 'pure' environmental policies such as quality or emission standards rather than accept imposition of specific fiscal, industrial or social policies.[8]

---

[7] Brian Wynne, 'Implementation of greenhouse gas reductions in the European Community: institutional and cultural factors', *Global Environmental Change*, 3,1: 101-28, 1993.
[8] *ibid.*

All these factors complicate the process of achieving agreement among EU countries. The problem is obviously much harder at the global level. The scientific and technical uncertainties, and the way they may be exploited by various interested parties, make it still more difficult. However, global environmental problems are best addressed through international coordination. If concrete action to counteract climate change has any chance of being taken in the near future, it will be easier for a small group of relatively homogeneous countries such as EU member states to take the lead. Such action may also demonstrate possible avenues for implementing collective agreements between states with different cultures and sensitivities. Hence, in addition to the general obligations on OECD countries, the importance of European climate policy for the world.

2.3 Regime development and the possible role of the EU

How might leading countries or groups interact with those that are more reluctant to take action? So far, the latter group has slowed down the former. We argue that the interaction can work in the opposite direction. An initially small coalition can take a lead and benefit by doing so; and such a coalition could grow, or otherwise stimulate others, through a combination of political and economic factors.

Initial commitment can lead to coalitions involving a small number of countries. This in itself yields inadequate environmental protection when the problem is global, and it may prove unstable if those countries perceive that they are incurring costs without making a significant impact on the problem or inducing others to act likewise. This is not necessarily the fate of such an initiative however.

First, it is by no means obvious in the case of climate change that implementing the European CO_2 stabilisation commitment need incur significant macro-economic costs (see Chapter 3). It may indeed help by providing added political incentives and rationales for policies which can anyway be beneficial: improving energy efficiency, removing subsidies and/ or reforming tax systems, and stimulating desirable technological developments and a market lead in lower CO_2 technologies. Furthermore, there exist various ways in which an initial coalition may expand. The fact and impact of political leadership (or at least sticking to aims declared under the UN Framework Convention) is obviously important, but there are also more direct economic mechanisms for creating and expanding the number of cooperating countries.

One way in which a core of committed countries could induce expansion is by means of self-financed transfers;[9] once some countries are determined to act, it may pay them to offer inducements to ensure wider participation. Such expanded coalitions can be profitable, but their stability requires some degree of commitment by a subset of the participating countries.

A second mechanism involves linking environmental policy to technological development and cooperation. Carraro and Siniscalco[10] show that the offer of technological cooperation can be used as a basis for building and expanding the group of countries. Stronger environmental policy can induce development of improved technologies; offering to share these and related developments with other countries if they join the regime can create incentives that make the regime profitable, stable and welfare-increasing. The benefit to those joining the regime is that they gain access to a stream of improved technologies and continuing R&D efforts. The benefit to those already within the regime is that their R&D efforts are enhanced by greater international cooperation, and the technologies produced gain preferential access to wider markets.[11]

The two mechanisms are different, but related. Both require initial leadership from a group of countries. In the latter case, however, the expansion can become self-sustaining even without invoking a sustained altruistic commitment on the part of the initial coalition.

---

[9] C. Carraro and D. Siniscalco, 'The international dimension of environment', *European Economic Review*, 36: 379-87, 1992; C. Carraro and D. Siniscalco, 'Environmental innovation policy and international competition', *Environmental and resource economics*, 2: 183-200, 1992.

[10] C. Carraro and D. Siniscalco, 'Policy co-ordination of sustainability: commitments, transfers and band-wagon effects, in I. Goldin and L.A. Winters (eds), *Sustainable Development*, CEPR and Cambridge University Press, 1994; C. Carraro, A. Lanza and A. Tudini, 'Technological change, technology transfer and the negotiation of international environmental agreements', in *International Environmental Affairs*, 1994.

[11] The formal conditions for the profitability, the stability, and the optimality of the joint negotiations on environment and technology cooperation are discussed in Carraro and Siniscalco (1994), *ibid*.

2.4 Conclusions

The implications of these various observations for the EU are clear. Ensuring that current commitments by OECD signatories (including the EU as a group) are met is an important political precondition for enabling the global climate negotiations to go further. In addition, because the situation in the EU has some similarities, on a smaller scale, with the kind of problems which need to be addressed at the global level, successful implementation of climate policy in the European Union is of particular importance.

Furthermore, we argue that if the EU countries do find an effective way of implementing their national and collective commitments, this may help to establish a nucleus for an emissions control regime that could expand over time. The expansion could occur partly due to the political effects such a development would have in encouraging other countries (initially, industrialised countries) to meet their commitments, and perhaps to associate with an EU scheme as a joint implementation of industrialised commitments under the Convention. But there are also more direct mechanisms by which such a coalition could expand, including involving developing countries, on the basis of economic interests. To an extent, the EU could offer financial inducements for such expansion on the basis of its commitments. More broadly, if the EU takes a leading position, coupled with careful design of technological innovation policy with particular attention to aspects of international cooperation, this could act as an incentive to bring other countries into an expanding emissions control regime.

Thus the fact that Europe accounts for 'only' about 15% of global CO_2 emissions is not a reason for avoiding significant action to implement the commitments made. The fact that Europe is the most promising candidate for establishing a leading and expanding coalition – and could benefit by doing so – suggests the reverse. For Europe and its industries to play such a role, however, Europe must first find a way of implementing its commitments in a way that is effective and efficient, that fosters innovation, and that can command political respect and support internally and internationally.

Chapter 3

The development of European climate policy

3.1 The stabilisation declaration of October 1990

After a degree of international scientific consensus about the threat of anthropogenic climate change had been established in 1985,[12] the issue began to acquire more political force. A sequence of international and intergovernmental meetings ensued, culminating in the Second World Climate Conference (SWCC) in November 1990, and during this period many countries declared national targets for CO_2 emissions. Just before the SWCC, the joint Council of EC Energy and Environment Ministers of member states met on 29 October, 1990 and agreed that:

> The European Community and member states assume that other leading countries undertake commitments along the lines mentioned above [ie. stabilisation of CO_2 emissions by the year 2000 at 1990 levels] and, acknowledging the targets identified by a number of member states ... are willing to take actions aiming at reaching stabilisation of the total CO_2 emissions by 2000 at the 1990 level in the Community as a whole.

The EC wanted to demonstrate a leading role at the SWCC. It was a political agreement, however, and not a legally binding EC Decision. The choice of the goal was not random. Several of the member states had already established national emissions reduction targets, as discussed below. On this basis, combined with assumptions about what the other member states might achieve, it appeared that the stabilisation of CO_2 emissions at 1990 levels by the year 2000 would be feasible.

---

[12] WMO/ICSU/UNEP Conference, Villach, Sept-Oct 1985.

3.2 Developing an EC strategy

Immediately after the stabilisation declaration was made, the Commission of the EC began to draft communications on a strategy to achieve this goal. In May 1991, a draft communication identified four major elements of a strategy: a regulatory approach; fiscal measures; burden sharing among member states; and the scope for complementary action at the national level. The idea of a Directive allocating emission targets for each member state was abandoned, and interest turned to harmonised regulatory and fiscal measures.

Just as the SWCC had precipitated an agreement, the UNCED 'Earth Summit' Conference at Rio in 1992 influenced the EC and its member states. The Commission attempted to develop a complete package of measures to be agreed together, which after various delays and amendments largely because of arguments about fiscal measures, was placed before the European Council just before the UNCED in June 1992. The proposed measures were:[13]

- A framework Directive on energy efficiency within the existing EC programme on 'Specific Actions for Vigorous Efficiency' (SAVE);
- A Decision on renewable energies – the ALTENER programme;
- A Directive on a combined carbon and energy tax;
- A Decision concerning a monitoring mechanism for CO_2 emissions.

Development of the draft Directive on the carbon/energy tax by the Commission proved to be particularly difficult. It was initially proposed to start in 1993 at the equivalent of 3 ECU per barrel of oil equivalent (boe) and rise by 1 ECU/ boe each year to a level of 10 ECU/boe in the year 2000. A compromise led to it being formulated as a tax half on energy and half on carbon content. Concerns about the impact of the tax on industrial competitiveness led to substantial exemptions for energy-intensive industries, and it was decided (partly because of electricity trade complications) that the tax should apply

---

[13] Details of the development of EC climate policy, including the specific elements and the legal context, are given in Nigel Haigh, ed, *Manual of Environmental Policy: the EC and Britain,* Institute of European Environmental Policy / Longman, London/Bonn, 1990 looseleaf (regularly updated); Chapter 14, Climate Change. A summary of developments to the end of 1992 is given in Pier Vellinga and Michael Grubb, eds, *Climate Change Policy in the European Community,* Royal Institute of International Affairs, London, 1993.

to electricity output rather than input fuels. All these changes weaken the impact on emissions, limiting the likely emission reductions to little more than 3% of the reference projection by 2000 even on the proposed schedule. In a further crucial change, the tax proposals were also made:

'... conditional on the introduction by other member countries of the OECD of a similar tax or of measures having an financial impact equivalent to the measures provided for in this Directive'.

Despite these changes, the Council still did not agree on the package of measures, most notably with continuing objections from some countries to the carbon/energy tax. Between the signing of the Framework Convention on Climate Change at Rio in 1992 and the ratification of the Convention in December 1993, the proposals were progressively amended in an attempt to get Council agreement – efforts which still continue, as discussed further below.

From the time of the initial proposal by the Commission to approval by Council, both the SAVE and the ALTENER programmes were reduced in scope and content. The enhanced SAVE programme was envisaged as a portmanteau for a series of Directives on energy efficiency measures that would contribute at least a quarter of the reductions necessary to achieve the stabilisation goal.[14] However, many important measures were dropped from the originally proposed package[15] and all detailed requirements were removed, leaving responsibility for programme content to the member states, on grounds of subsidiarity. The impact of the programme as adopted will be far weaker than originally envisaged.

Similarly, with the ALTENER programme, proposals were made during its development for a large and well-financed programme for implementing renewable energy investments in the Community. The final Decision on renewable energy (ALTENER) came into force on 1 January 1993 and

---

[14] Nigel Haigh, ed, *Manual of Environmental Policy*, Institute for European Environmental Policy / Longman, London/Bonn: Release 4, p. 14-5-2, 1994.
[15] For example, a measure to ensure regular inspection of cars was dropped and energy audit requirements weakened.

contained specific targets, but without substantive tools for implementation and with Community funding of ECU40 million for the first three years – a small fraction of the cost of a single large power station.

The Decision for a Monitoring Mechanism of EC CO_2 and other greenhouse gas emissions was adopted by Council on 24 June 1993. This Decision requires member states to publish and implement national programmes for limiting CO_2 emissions, including inventories and emission projections to the year 2000 (see box). From these reports, the Commission has to calculate an inventory and projections for the EC as a whole. The results of the first compilation are discussed below.

The carbon/energy tax proposal continued to be opposed by several actors – within the Commission, industry and some member states. In the Council, France opposed the tax because they favour a carbon tax, given their reliance on nuclear energy. The less developed countries have opposed the introduction of a tax, since they feel it would be a barrier to development, and the UK opposed it on the grounds that taxation should not be decided at EC level.

During the process leading to ratification of the UN Climate Convention, six of the countries, led by The Netherlands and Germany, attempted to link the EC ratification to the adoption of the EC carbon/energy tax, claiming that the 1990-2000 stabilisation goal could only be reached through the adoption of the tax. However, the UK refused to change its position, while countries like Spain began to worry that if the EC did not ratify the Convention, they would have difficulties in fulfilling their commitments under the Convention. In December 1993, wording was agreed that allowed the Community to ratify the UN Convention whilst minimising the loss of face for proponents of the tax.

The net effect of these changes is that European-level policy has been rendered almost impotent. The measures under SAVE and ALTENER are each likely to reduce CO_2 emissions in the year 2000 by less than 1%. Efforts are continuing towards some form of tax agreement, but as discussed below even if some compromise is reached it is unlikely to have significant impact on emissions, at least by the year 2000. Excepting this, and the technology measures taken under the Community's R&D programmes, the Community's role has been relegated to monitoring what member states choose to do unilaterally, and projecting the consequences.

Decision for a Monitoring Mechanism of CO2 and other greenhouse gas emissions

(Extracts from European Council of Ministers Decision, 24 June 1993)

' .. Whereas on the signing of the [UN Climate] Convention the Community and its Member States reaffirmed the objective of stabilisation of CO2 emissions by 2000 at 1990 levels in the Community as a whole ..

2.1 The Member States shall devise, publish, and implement national programmes for limiting their anthropogenic emissions of CO2 in order to contribute

(i) to the stabilisation of CO2 emissions by 2000 at 1990 levels in the Community as a whole, assuming that other leading countries undertake commitments along similar lines, and on the understanding that Member States which start from a relatively low levels of energy consumption .. are entitled to have CO2 targets and/or strategies corresponding to their economic and social development

(ii) the fulfilment of the commitment relating to the limitation of CO2 emissions in the UN Framework Convention on Climate Change by the Community as a whole throughth action by the Community and Member States, within their respective competences.

2.2 Each Member State shall, at the latest from the first updating, include in its national plan: .. details of national policies and measures .. and trajectories for its national CO_2 emissions between 1994 and 2000 ...

5.3 The Commission shall evaluate the national programmes, in order to assess whether progress in the Community as a whole is sufficient to ensure fulfilment of the commitments ..

5.4 The Commission shall report to the Council and the European Parliament the results of its evaluation within six months of the reception of the national programme

6. After the first evaluation .., the Commission shall annually assess in consultation with the Member States whether progress in the Community as a whole is sufficient to ensure that the Community is on course to fulfil the commitments ...'

Source: Decision 93/389/EEC, OJ No. 11 167/31

Figure 3.1 Distribution of CO_2 emissions in EU-12 and Accession countries, 1993

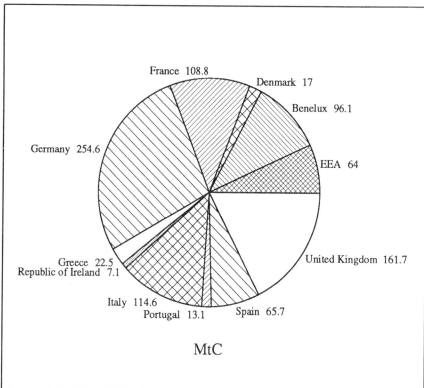

Source: derived from BP Statistical Review of World Energy, 1994, using emission factors from M.Grubb et al, *Energy Policies and the Greenhouse Effect: Volume II*, RIIA, London (Appendix II).

3.3 Current projections and country positions

The prospects for achieving the European CO_2 stabilisation goal therefore need to be understood primarily in terms of the prospects for national emissions. Figure 3.1 shows the distribution of fossil fuel CO_2 emissions in 1993 amongst the main countries and sub-groups in the European Union and Accession countries. No country dominates, but clearly the German position, with 27.5% of the total, is especially important and the biggest four countries together accounted for nearly two thirds of the total in 1993.

Table 3.1 National CO$_2$ emissions and projections for EU-12 countries

	1990 CO$_2$ Emissions (million tonnes)		
	Total	per capita	Declared national emission target/goal
Belgium	112	11.2	5% reduction from 1990 by 2000
Denmark	53	9.9	5% reduction from 1990 by 2000
Germany	1005	13.0	25–30 % reduction from 1987 by 2005
Greece	74	7.4	25% increase by 2000 from 1990
France	366	6.5	13% increase by 2000 from 1990
Ireland	31	8.8	20% increase by 2000 from 1990
Italy	402	6.9	Stabilisation on 1990 level by 2000
Luxembourg	12.5	35.0	Stabilisation on 1990 level by 2000 (at latest)
Netherlands	157	12.2	3–5% reduction from 1989/90 by 2000
Portugal	40	4.1	30–40% increase by 2000 from 1990
Spain	211	5.4	25% increase by 2000 from 1990
UK	579	10.2	return to 1990 level by 2000

Most member states have declared emission goals relative to the year 1990. National CO$_2$ emissions in 1990, and declared goals for the year 2000, are summarised in Table 3.1 for each of the EU-12 countries.[16]

By adding up these national targets, it is possible to calculate the implied Union emissions for the year 2000, except for Germany which is the only EU-12 country not to have declared a target for the year 2000. As indicated in Table 3.2, the German position is crucial. Germany's CO$_2$ emissions decreased by 14% between 1987 and 1993 – most of this since 1990 – but this reduction was due to the economic collapse in the former East Germany; there was no emissions reduction in the former West Germany. If Germany achieves a 12.5% reduction from 1990 levels in the year 2000 – which is not unreasonable given the present reductions and the goal of a 25% reduction by 2005 – and all other countries meet their declared goals, then EU-12 emissions would indeed roughly be at 1990 levels.

---

[16] Note that these are goals, not projections. The goals are compared against projections by the European Commission in Vellinga and Grubb, *Climate Change Policy in the European Community*, op.cit, Table 1. The Commission's projections frequently vary from those of the member states, and there are no institutional mechanisms for exploring the reasons for differences and achieving consistency. For example, the Spanish official national emission target is a substantial reduction compared with their own 'business as usual' projections, but a fractional increase on the Commission's projection.

Table 3.2 Total CO_2 emissions and year 2000 targets for the EU-12 (million tonnes)

	1990 emissions	2000 targets
EU-11 (excluding Germany)	2037	2165
Germany with 12.5% reduction in 2000	1005	879
Total (EU-12)	3042	3044

However, the reality is more complex. The European Commission issued a first evaluation of the existing national programmes under the monitoring mechanism of EC CO_2 and other greenhouse gas emissions in March 1994.[17] The Commission noted that only two countries – The Netherlands and Denmark – had completed a full and specific programme to achieve their target. Germany continued to decline specifying a CO_2 emission projection for the year 2000, and many other national reports contained inadequate detail.

Various indicators suggest that several countries are having difficulties meeting their targets. Many of the member states have submitted plans that indicate the *technical possibility* of meeting their targets and discuss potential policy measures. But several do not detail measures in place, or necessarily agreed by the government, for achieving these goals, and/or do not contain plausible analyses of the impact of measures being implemented upon national emissions.

This is not the case for all countries. In a few, developments and projections suggest that targets may be over-achieved.[18] But overall, the Commission's March 1994 review of national reports[19] stated that 'existing national strategies to cut emissions of CO_2 indicate that the Community is unlikely to meet its

---

[17] COM(94) under Council Decision 93/389/EEC, 10 March 1994

[18] For example, in the UK, the impact of electricity privatisation coupled with refusal to protect the coal industry, prolonged recession, the introduction of VAT on domestic energy and increased taxes on petrol, together seem likely to result in emissions in 2000 being below 1990 CO_2 levels, the official target. The official goal for some of the Southern European countries allowed considerable – and perhaps excessive – headroom for increased emissions.

[19] COM(94)67 final.

target of stabilising emissions at 1990 levels by 2000.'[20] The Report adds: 'For these reasons more consideration should be given to other parts of Community strategy, in particular the CO_2/energy tax proposal, which is more cost-effective.'

This does not mean that the goal is either impossible, or costly, to achieve. On the contrary, technical assessments indicate that it should be possible to achieve the EC CO_2 stabilisation target at little or no cost.[21] The difficulty lies in implementing policies that can exploit effectively the technical opportunities; and the primary problem is one of finding the political will, and policy sophistication, required to achieve agreement upon such policies within member governments across different departments and interests.

Furthermore, few of the national plans include strategies or convincing projections beyond the year 2000. The Dutch and Danish projections and goals go to the year 2005. The German target, as noted, is for a 25% reduction by that date; but without the 'gift' of additional reductions from East Germany, and with constraints arising from commitments on German coal and trends in transport and other sectors, this seems unlikely to be achieved. At present it seems that the prospect beyond the year 2000, far from entailing further reductions in line with the changes that would be needed to start meeting the objectives of the UN Climate Convention, is that CO_2 emissions may start rising again.

3.4 Further development of the carbon/energy tax

The ratification of the UN Climate Convention did not end efforts to find an acceptable way of implementing a carbon/energy tax in the European Union. The Greek Presidency of January to June 1994 presented new and substantially different proposals. Instead of an automatic annual increase in the tax rate, it would initially be held at the low (US$3/bbl) level pending a review of progress

---

[20] For a summary see Financial Times *EC Energy Monthly*, April 1994.
[21] Commission of the European Communities, *Cost effective analysis of CO_2 reduction options*, Synthesis report / country reports, by COHERENCE for the CEC, DG XII (Research), Brussels, 1991; also see various papers in *European Economy*, Special edition No. 1, CEC, DG II (Economic Affairs), Brussels, 1991. Developments in East Germany, trends in electricity generation, and revised economic growth forecasts suggest that the target should be easier to achieve that at the time of these initial assessments.

after three years, and it would not be implemented in all countries. In response to the misgivings of Greece, Spain, Portugal and Ireland, there was greater emphasis on 'equitable burden-sharing', for which an index was proposed that would have the effect of exempting these countries until their CO_2 emissions reached a certain level (to be negotiated). There could be allowances for countries that had already introduced national carbon/energy or related taxes. Furthermore, the Greek proposal would exempt energy intensive industries from the tax according to the extent to which energy costs are reflected in the end product value. Given all of these exceptions, and the low level of the tax, the Greek proposal for a tax would not lead to significant emissions reduction; but agreement still could not be obtained in the Council.

Subsequently, under the German Presidency of July to December 1994, the focus shifted towards using the existing system of excise duties so as to avoid objections to creating a 'new' form of EC-level taxation. Instead, it was proposed that the scope of current excise taxes on motor fuels could be extended to some other fuels. The Germany Presidency proposed this in a package with a statement recognising the importance of EC carbon/energy taxes in principle. But although some member states already have some form of carbon/energy tax, fundamental opposition to the idea of a tax agreed at EC level remains. In particular the UK, but also Spain and Portugal, basically disagree with providing the EC with the competence to develop tax proposals because of sovereignty implications. The September 1994 Council failed yet again to reach agreement. Whether or not an agreement is reached under the German or subsequent French Presidencies, it now seems clear that it will bear little or no resemblance to the original proposal, and will have little impact on emissions, at least by the year 2000.

Yet there is no doubt that energy pricing is important in determining energy choices; consequently, so are price-based measures such as carbon taxes. An element of carbon tax really is unavoidable as a strategic component in seeking to alter long-term trends in energy consumption and fuel choice. But political obstacles have, as we have seen, so far blocked all attempts to get meaningful agreement on EC-wide taxes; after more than three years of intensive political effort, and despite the strongest backing from Europe's most powerful state, little progress has been made. The impasse has also afflicted other non-tax measures that could also make important contributions to limiting emissions.

Yet the increasing integration of the European economy also makes it very difficult for countries to consider unilateral action.

The experience reveals several reasons for these problems. Most tax measures are politically difficult. Political opposition, arising from both industry and consumer / social protection groups, have proved formidable obstacles to the carbon/energy tax. Agreement at the EC level has been further complicated by the different situations in each country. Furthermore, tax measures require unanimous agreement in the EC Council of Finance Ministers, placing the central onus upon institutions that generally have little interest in, or concern about, the climate issue. To make progress, the benefits of limiting national emissions need to be made more tangible to these parties.

3.4 Conclusions

European Community measures to tackle climate change have not fared well. The SAVE and ALTENER Programmes are much weaker than originally proposed and any near-term agreement on a carbon/energy tax seems likely to be on a much weaker proposal than originally conceived. Sovereignty and international competitiveness arguments have so far blocked progress on development of an EC strategy to reduce CO_2 emissions.

Thus, EC policy at the end of 1994 amounts to little more than a policy of monitoring progress towards the targets that many countries had already unilaterally adopted by 1990, backed up by a technology programme and promises that if progress proves inadequate the Commission should consider the issue again. Present evidence from EU member states suggests that the goal of stabilisation of CO_2 emissions at 1990 levels by the year 2000 will be difficult to reach, since many individual member states are not implementing measures sufficient to meet their own targets; and current measures are clearly inadequate to achieve goals of longer-term reduction.

Perhaps most disturbing, it is arguable that the Union and its member states have actually impeded each other's efforts to address climate change: that so far, the sum of responses may be less than the parts might have individually achieved. Member states, as we have seen, initially asked the Commission to develop proposals for meeting the CO_2 target and then blocked or neutered virtually all the substantive Commission initiatives on climate policy. Neither has the Commission gained any significant powers to make member states stick to their own unilateral promises on emission targets. In turn, the integration of the European economy has made it more difficult for countries to act unilaterally, and has also diverted political effort and argument from domestic initiatives to EC-level negotiations and expectations. Industry in Germany and other central EU countries, for example, argued powerfully against any serious consideration of unilateral fiscal measures on the grounds that they should be taken only when coordinated at the EC level; industry (and government) in the UK and some others rejected the EC proposals on precisely the opposite grounds, that such important measures were a matter for national sovereignty.

Thus, if the EC is to play any constructive role in tackling climate change; if the EC commitment is to be met; and certainly if longer-term reductions are to be achieved, fresh approaches need to be considered. Europe needs to find approaches that induce member states to achieve their stated individual emissions targets, and to negotiate more seriously the harmonisation of EC-wide measures. We consider such an approach below. First, we turn to one other important element: the expansion of the Union.

Chapter 4

European Union expansion and the Accession countries

Another important element in EU climate policy up to the year 2000 is the expansion of the Union, and its relationship with other closely affiliated countries. Expansion to include several countries of the European Economic Area raises the question of how these countries' emission targets and policies relate to those developed by the former European Community.

The emission situation and targets for the four countries in question are shown in Table 4.1. Together, these four countries account for an additional 200 million tonnes of CO_2, about 6 per cent of the EU-12 total. In terms of per capita emissions, apart from Finland they are well below the EU-12 average.

At the time of writing, referenda in Austria and Finland have confirmed that they will join the EU; the other two are awaited. The possibility that one or both of these countries may not join highlights another important aspect of EU climate policy: to what extent can or should policy seek to be flexible enough to incorporate closely related countries into a framework wider than just current Union members? Or, conversely, is there a danger that climate policy could become so complex and specific as to add further to the complications and difficulties of future Union expansion? Irrespective of the outcome of the Scandinavian referenda, for a strategic issue like climate change this is an important aspect to consider, with clear bearing upon the potential for the EU to be a nucleus for wider solutions as discussed below.

The current expansion of the Union – whether or not it includes Sweden and/or Norway – has at least two important implications for European climate policy. First, all these countries joining have strong environmental sensitivities. They were early members and often strong proponents of many international environmental agreements,[22] and were among the first countries to ratify the UN Climate Convention. They all consider fiscal instruments to be an important part of climate policy, and support, in principle, changes in energy

---

[22] For data, see *Green Globe Yearbook 1994*, Fridtjof Nansen Institute, Oslo, Norway, 1994.

Table 4.1 National CO$_2$ emissions and projections for the Accession countries

	1990 CO$_2$ Emissions (million tonnes)		
	Total	per capita	Declared national emission target/goal
Austria	58.5	7.5	20% reduction from 1988 by 2005
Finland	55.0	11.0	Stabilisation in second part of 1990s
Norway	32.4	7.6	Stabilisation at 1989 levels by 2000
Sweden	55.8	6.5	Stabilisation at 1990 levels by 2000; reductions thereafter

taxation to internalise environmental damages. Along with Denmark, the
Nordic countries also form the only countries to have implemented carbon/
energy taxes justified largely on environmental grounds (at rates equivalent
to up to $200 per tonne of carbon),[23] though these take widely different forms
in the different countries. They will add weight to environmental sentiment in
the European Council, both in general and specifically in the context of energy
and climate policy. The fact that concern about the perceived weakness of EC
environmental policy has played a significant role in anti-EU sentiment in
some of these countries (as in the initial Danish rejection of the Maastricht
Treaty) may further heighten sensitivities on environmental topics.

The second impact is that explicitly on the EC emission goal. The national
targets summarised in Table 4.1 suggest that the accession of these countries
will make little difference to the EC stabilisation commitment, since all are
nominally committed to that or a very similar goal. But the prospects for
achieving these targets are not good and the national authorities have put on
record important caveats to these general commitments:

- The Finnish government has indicated that it will be practically impossible
  for Finland to achieve the stabilisation target, and consequently has confined
  its ambition to halt the increase in energy-related CO$_2$ emissions by the
  end of the 1990s.
- Austrian emissions increased by 14% from 1988 to 1991; the Austrian
  report to the Climate Convention includes scenarios with substantial

---

[23] The nominal rates are highest in Sweden and Norway, but it is difficult to distinguish
nominal from 'effective' rates due to substantial exemptions and differences in tax structures.

emission increases, and a recent non-governmental report estimates a possible increase by 23% by 2005.[24]

• The Norwegian Ministry of Environment recently projected a 13% increase in emissions by 2000 without much more stringent policies, and the stabilisation objective is stated to be preliminary and to be re-evaluated in the light of new developments.[25]

• Swedish forecasts suggest that their emissions could increase 10% by 2000, though measures proposed are deemed adequate to stabilise by 2000; thereafter, implementing the proposed nuclear phase-out would require much tougher measures to suppress emissions.

The obstacles to more effective emissions control originate in the energy structures of these countries. With the partial exception of Finland, they all start from a base of very low per capita emissions, reflecting the dominance of non-fossil sources (mainly hydro and nuclear) for power generation. The main emissions are from transport and industrial fuel use (which is itself already quite efficient). Thus unlike much of the EU-12, they do not have scope for offsetting rising transport sector emissions by reducing emissions from electricity production. Furthermore, they are all small, open economies, particularly vulnerable to concerns about international competitiveness and industrial emigration. Thus, they face major difficulties in living up to their promises on a purely domestic basis, and are looking increasingly to forms of international collaboration as a way forward.

Thus, although these are not big emitters, expanding the Community's collective stabilisation goal to include the new members is likely to add fractionally to the overall effort required to achieve the collective EC target. The alternative, of simply insisting that the EC's goal only applies to the original Twelve, appears legally questionable,[26] and would introduce a barrier

---

[24] Climate Network Europe, 'Independent NGO Evaluations of National Plans for Climate Change Mitigation', Climate Network Europe, Brussels, 1994, pp 5-6.

[25] For elaboration see Anne-Kristin Sydnes, 'Norwegian climate policy: environmental idealism and economic realism', forthcoming in Jill Jaeger and Timothy O'Riordan, eds, *The Politics of Climate Change in Europe*, Routledge, 1995.

[26] The Community ratified the UN Climate Convention as a legal entity; from 1995 that entity will include the new entrants.

to principles of economic integration and efficient sharing of the Community's international commitments. But if it is correct that expansion will require slightly greater efforts in the rest of the Union, because the accession countries cannot meet their declared targets, they may be expected to contribute in some way to efforts in the rest of the EU – again raising the question of what implementation strategy can allow for such burden-sharing.

Chapter 5

A fresh approach

5.1 The policy dilemma

While the EC's commitment to stabilisation is an important step, the real difficulties, as so often in environmental policy, lie in the implementation. It is especially difficult for a pervasive and crucial industry like energy, in the complex and evolving political make-up of the European Union and Community institutions.

For some of the relevant measures, there clearly is a strong case for harmonising action across the EU. Examples include efficiency standards on tradeable goods, and policies for demonstrating new technologies and supporting development of an adequate industrial manufacturing base for them. There are reasons also for seeking greater harmonisation of fiscal measures, though as noted in Chapter 3 the variations in existing tax structures and political attitudes, have so far proved powerful obstacles.

Yet, having other measures established at the Union level makes little sense: building insulation standards are very relevant, for example, but no one trades buildings, and even if they did they would hardly want the same standards in Portugal as in Denmark. Indeed, all kinds of issues – the form of utility regulation, VAT distortions, transport policy, etc. – affect CO_2 emissions, and it is clearly not realistic to suppose that all of these can or should be coordinated across the EU as part of the stabilisation policy. This, combined with the broader political and cultural differences between member states noted in Chapter 2, makes it clear that the key energy policy decisions required to stabilise emissions cannot and should not all be taken centrally. Yet the goal remains inherently a collective one, and the potential economic, environmental and political benefits of cooperating among member states remain very large.

5.2 National emission targets

An opposite approach to that of centralised development of energy policies for limiting CO_2 is simply to negotiate CO_2 emission targets for each member state, such that the total adds up to the stabilisation goal. This has the political advantage of being a very simple and well understood approach, which leaves the specific energy policy decisions required to meet the emission targets to the member states.

In one sense, Europe is already some way down this road, because as noted in Chapter 3, most member states have already declared national emission targets for the year 2000, and taking these targets at face value they add up to a total close to EU stabilisation. But there are several problems with relying on these national targets, backed up by the Monitoring Decision. Most importantly, as noted in Chapters 3 and 4, countries differ in the seriousness accorded to their targets and the prospects for achieving them. It is unlikely that all national targets will be met, and there is no direct incentive – other than political face-saving – for them to make greater efforts to achieve their declared goal.[27] The Commission, through the Monitoring Decision, can alert when national policies are not adequate to meet the declared emission goals – and it has done so. But what steps then are open?

Attempting to convert existing national targets into legal commitments – which would be somewhat analogous to the Large Combustion Plant Directive for sulphur dioxide emissions – faces a number of problems. These arise principally because CO_2 is a much more fundamental issue, associated with industrial structure rather than end-of-pipe cleanup, offering less scope for closely targeted short-term reductions in national emissions. The current difficulties indicate that many underestimated the political difficulty in devising and implementing the required policies, and in particular the problems of getting different institutions – such as departments of industry and finance

---

[27] Relying on political embarrassment is scarcely likely to be adequate for an issue like CO_2, and there is an additional institutional problem: primary responsibility for many of the energy and fiscal policies that are relevant to CO_2 emissions lie with Ministries of Industry and Finance, respectively; but it is Environment and Foreign Ministries that are likely to be most embarrassed by failure to meet an environmental target.

that have little direct interest in the climate issue – to agree on policies. They also underestimated the inherent uncertainties involved.

Thus, in the context of CO_2 the approach of setting fixed national targets appears insufficiently flexible. The process of setting such targets is so fraught and difficult politically that the prospects for revising the distribution of emissions between countries, if this proves justified in the light of national trends and experience, are negligible. In any attempt to negotiate legally binding and fixed targets, this same factor creates a very powerful incentive on all the negotiating parties to ensure that they get the highest possible emission target, with maximum headroom for uncertainty in emissions.

Nor is such a system efficient, because the targets might require more difficult or high cost measures in one country whilst simpler abatement opportunities elsewhere remain unexploited – an argument that has already been used to oppose such a system. For example, as noted in Chapter 4 the Nordic countries start from positions which make it inherently more difficult for them to achieve stabilisation goals and they may question increasingly whether it is sensible for them to pursue reductions domestically that can be achieved more cheaply elsewhere in the Union. But an agreement on fixed and binding emission targets would give other countries, that could over-achieve their target, no incentive to do so. Indeed, at present the incentive is the opposite – any country which finds its target easier to achieve than expected has an incentive to hide the fact, and/or delay policies to limit emissions further – and there are indications that this is already happening in certain EU countries.

For all these reasons, a system of fixed and binding emission targets for each member state would not enable the EU or its member states to realise most of the potential benefits of being a Union. In the aftermath of the October 1990 declaration, the approach was cursorily examined and politically rejected.

5.3 Tradeable emission quotas

A key need is therefore for a system that encourages member states to take more seriously the contribution of national emission commitments to the collective European target: one that embodies penalties for exceeding initially agreed emission targets, and rewards overachievement.[28] An extension of the target-sharing approach offers a way of achieving this, and may provide a desirable way of resolving the central dilemmas in seeking to implement the Community's CO_2 emission commitment. National emission targets, or 'quotas' could be used, but with the critical distinction that the member states, or their industries, would be free to 'trade' them with others.

In other words, the Union could create 'emission quotas' for carbon totalling the already agreed level (i.e. stabilisation in 2000 at the 1990 level) and negotiate an initial division, but these would not form fixed targets. Participants would undertake to ensure that their emissions in the target year (2000) do not exceed the quotas they hold in that year. If their initially agreed quota allocation proves insufficient, they would have to obtain, from other member states, additional quotas. Thus, some countries could let their emissions exceed their initial allocation if they obtain quotas from others whose abatement efforts leave them with spare – and who would thus be rewarded accordingly.

Thus, there would be an incentive on all countries to minimise emissions (either to minimise the payment for quotas or to maximise the revenue from selling them) to a degree consistent with achieving agreed Union goals at least cost. National bureaucracies – and in particular finance ministries – would be faced directly with the fact that CO_2 emissions involve a tangible cost, and could thus balance internally the benefits of constraint against more traditional energy policy goals. Ultimately the 'price' of such quotas should settle at a point which reflects the least costly way of meeting the stabilisation target anywhere in the Union. The efficiency benefits could be considerable; one study suggests that the costs of an approach which allows such inter-

---

[28] One option for achieving this could be to negotiate a system of explicit fines (and reimbursements) that would apply according to the degree of under (and over) achievement of agreed targets. One problem with this could be that the rate of the 'fine' – which would of course have to be agreed between all member states – would tend to be driven down to the lowest common denominator by the country that perceived the greatest difficulty in meeting its target; and thereby may prove inadequate overall for ensuring collective stabilisation. How to spend any money raised – or to fund reimbursements – would add further complications.

country flexibility could be just one-fiftieth of the costs involved if each country were bound to stabilising CO_2 emissions individually.[29]

Such a system ensures that the collective goal of stabilising total emissions is attained, because this is established by the total number of quotas issued. But it is much more flexible than the allocation of fixed targets. It is also fully consistent with the 'polluter pays' principle by ensuring that increases in emissions above the agreed initial quotas are paid for, and constraint is rewarded. The fact that the EU total would be fixed, and the terms of national commitments clear, would also reduce the uncertainties currently pervading relevant industries.

In such a system, governments would retain control over the policies used to limit emissions, but components could be adjusted for mutual benefit under the broad thrust of overall EU harmonisation. The Commission could still promote European-wide components to energy and climate policy; indeed, governments would probably be more receptive to them because the benefit of limiting emissions would be more tangible. For example, reform of tax systems to encourage emission reductions over time in concert with other policy goals could continue to be promoted in concert with a tradeable quota system. But these and associated measures could evolve as a strategic component of policy from the 'bottom up', along with a range of other developments, rather than tax harmonisation being a prime focus of dispute as part of the stabilization strategy for the year 2000.

The approach would thus provide an efficient and feasible way of meeting the declared goal, whilst being consistent with two major policy principles enunciated by the EU and agreed by the member states: the subsidiarity principle, by devolving the detailed energy policy decision making as far as is consistent with Union objectives; and the polluter pays principle.

Could a tradeable quota system work? How might it work? How might decisions on the initial allocation be made? And how might such a system relate to other policy measures, such as fiscal instruments? The next chapter sets out some of the issues and options. Before turning to these specific aspects,

---

[29] Scott Barret, 'Reaching a CO_2 emission limitation agreement for the Community: Implications for equity and cost-effectiveness', CEC DG II, Brussels, 1992.

it is useful to note the flexibility that such an approach confers with respect to enlargement of the Union, and broader participation.

5.4 Union enlargement and system expansion

Enlargement of the Union highlights another potential benefit to developing a system of tradeable emission quotas for implementing collective Union emission goals. The countries seeking to join in 1995 have several common features.

First, they have all established official objectives for controlling CO_2 emissions; even though the form, strength and explicitness of the commitment varies, all have taken a political commitment, as a minimum, to stabilize these emissions by the year 2000.

Second, these commitments were formulated and promulgated in the early period of 'climate euphoria', but the energy structures of these countries – starting mostly from low per capita emissions and with little or no use of carbon fuels in electricity production – make it unusually difficult for them to achieve stabilisation. Politicians and civil servants in the four countries (as in some others) have now reached the stage where they have explored a range of options and reached the conclusion that it will be very difficult, verging on the impossible, to live up to their commitments.

Third, most of these governments have a high environmental profile and publics sensitive to environmental issues, and/or they have taken a particularly activist role in the international climate discussions. Consequently, they can hardly afford to recognise openly a failure to meet their commitments, and are thus searching for a way out of their dilemma.

All of these countries probably face relatively high domestic costs for CO_2 stabilisation, but are sufficiently rich and committed to be willing to consider financing abatement measures elsewhere, if they could somehow gain credit for this as contributing to their domestic targets. This is why the government with the most painful dilemma – the Norwegian under Prime Minister Brundtland, at home often referred to as the 'world environment minister' – has sought strenuously to promote the idea of 'joint implementation' of emission commitments between countries as part of the nascent climate regime.

The great majority of this diplomatic effort has focused upon the idea of funding for specific abatement projects in countries of the developing world and perhaps Central/Eastern Europe. But these proposals have faced severe opposition. There are practical problems associated with implementing such projects and monitoring the actual emission savings. There are political problems arising from ambiguities about who would really control and guarantee such projects, and the feeling that it simply represents a way of letting the rich world buy its way out of the obligation to get its own emissions under control. Despite all the effort, the developing world continues strongly to resist the idea, and it seems unlikely that a generally acceptable scheme for such joint implementation can contribute significantly towards achieving emission goals under the Convention for the year 2000.

Thus, joining the joint stabilisation commitment of the European Union and contributing to its implementation through a tradeable quota regime could turn out to be the only realistic solution for these countries. The proposed quota system would allow these governments to buy emission entitlements of a sufficient quantity to offset the forecast increases. If each of these countries had initial quota allocations corresponding to their declared emission objectives, for example, all four would probably emerge as active buyers of CO_2 quotas. The four applicants, in short, would be able to circumvent the exceptionally high cost of constraints in their own domestic energy systems by using their relative economic strength to pay other EU countries to adopt greater emission reductions. They would thus be able to avoid major political embarrassment by claiming, correctly, that they were helping to support and finance the larger and globally more important objective of stabilising total EU emissions.

The uncertainties about the scope of expansion raised by the impending (at the time of going to press) referenda highlight another feature of such a tradeable quota regime. There is no fundamental reason why it should be restricted to members of the EU. If, for example, Norway did not join the Union, it could still be of mutual benefit to allow Norway to join such a system. This would ensure bilateral joint stabilisation between Norway and the EU – or one of its member states – which would be of benefit to both parties and, in terms of its impact on the global processes, clearly preferable to Norway simply having to admit failure to achieve its (usually demanding) goal.

Such an extension of a European climate regime could, in principle, be offered to other prospective members and conceivably even to governments with no intention ever to join the Union. If the system did prove effective, the same approach could be applied to any governments that have undertaken stabilisation or related commitments and are prepared to back this up with real commitment of resources: if their baseline objective, or other negotiated quota, is acceptable, they can be allowed into the EU emissions 'bubble' if they are willing to take on the required reporting, monitoring and enforcement obligations.

Along with these obligations on participants could come other benefits – for example, greater collaboration on and access to the benefits of technological innovation associated with CO_2 constraints. Thus, such an approach could ultimately lay the basis for an expanding coalition of countries along the lines discussed in Chapter 2. But this, at present, is to run too far ahead. What matters at present is to establish whether such a system is feasible in the real world of European politics, and if so, what practical issues need to be addressed.

Chapter 6

Aspects of system design

Whilst the overall concept of a tradeable quota system among European countries is simple, developing a realistic and workable system would raise many complex technical, political and legal questions. This section discusses some of the technical and political issues; we leave the question of how such a system could relate to, and be developed within, the existing body of Community law for more specialised researchers to consider. We offer observations and options – not prescriptions – for there are a number of choices that could be made and many would be best resolved through further governmental analysis and negotiation.

6.1 Quota allocation

Politically, the most difficult issue is likely to be the initial allocation of quotas, since this determines the net financial gains and losses.

This is not in reality an issue that is unique to tradeable quotas. Any scheme for achieving emissions stabilisation in Europe implies some distribution of the effort involved, and this has been recognised in the declarations on the subject surrounding the carbon/energy tax proposals, and the development of the Community's Cohesion funds. The idea of the richer members of the Union being willing either to take more action, or to help support some of the poorer members in their actions towards collective Union policy, is nothing new. The allocation of tradeable emission quotas offers a particular – and particularly efficient – way of implementing this in the context of efforts to limit CO_2 emissions.

Such negotiations on burden-sharing in the Community have never been easy, and quota allocation would be no exception; but, as with budgetary disputes, resolution should be possible if there is a collective will to move forward.

Two broad approaches could be taken towards allocation. Given that positions have already been taken, goals declared, and emission projections made, allocation proposals could be partly or wholly based upon existing national targets. Quotas allocated according to declared national goals would be a way of ensuring that countries took their commitments seriously; it would in effect be asking them to 'put their money where their mouth is'. If all countries met the declared goals as set out in Table 3.1, the stabilisation objective would be achieved, with little or no transfers. In practice, as we have seen, this is unlikely to be the case. If the quota allocation were closely aligned to current national targets, countries which are richer and which have adopted more ambitious goals would tend to end up paying those with less ambitious initial goals. To the extent that poorer members of the Union have already adopted less ambitious goals, in effect the richer countries would be paying them to undertake greater abatement than they otherwise would.

An obstacle to this is that ambition in declaring national targets has not always correlated with wealth, and different countries have taken different approaches to the meaning of their declarations, as indicated in Chapter 3. Some were declared after considerable analysis as commitments to be met; others declared targets as general planning goals, and perhaps not commitments against which they expected to be held accountable. Negotiations on quota allocations would have the benefit of clarifying just how serious the different countries are about their goals, but the feasibility of allocation simply on the basis of declared goals remains to be determined.

Another approach to allocation would be to try to develop a crude formula for quota allocations. If it can be done, this has the twin advantage that (a) it reduces the scope for special pleading by each country – the negotiations would focus more upon what are reasonable criteria for allocation, though still inevitably influenced heavily by what the outcome would imply for national interests – and (b) it would provide a much simpler basis for expansion. A number of formulae, and variables which could feed into an allocation formula, can be considered. A particularly simple and appealing one, which still leaves some room for negotiation to avoid politically infeasible transfers, would be to make quota allocations a weighted combination of population and emissions in a past year or representative period. The higher the population weighting, the greater the benefit to countries with lower than average per capita emissions

– which tend to be the less developed members of the Union. The greater the weight accorded to previous emission levels, the more reductions from present levels will be rewarded, and increases penalised.

In practice, political realities are likely to dictate that any quota allocation established for the year 2000 is heavily influenced by current commitments, but an allocation formula – perhaps with exceptions for special circumstances or otherwise modified to reflect existing commitments – could be considered if such a system were to be extended.

Indeed, the possibility of extension beyong the year 2000 is in itself a final and important factor which may modify the political difficulties involved in allocation. The Convention's Objective implies long term global emission reductions, and the EC's emission declaration itself refers to 'stabilization' at 1990 levels, implying continuation of the commitment beyond the year 2000. Countries would be negotiating against a background of this long-term expectation. This possibility has implications for the politics of negotiating allocations.[30]

For example, if a tradeable quota system is established as an effective and efficient mechanism, it may prove desirable to extend the system to address subsequent emission goals with further rounds of allocation. A country which ends up with a large surplus of quotas in the year 2000 would be under pressure to accept a lower allocation in subsequent rounds; and if the situation arises because they held out for an unreasonably high allocation, based on implausibly high projections of CO_2 emissions, the credibility of their negotiation position for subsequent rounds would be seriously weakened. This principle, in fact, has relevance whether or not subsequent control is implemented by extending a tradeable quota system; it is intrinsic to the politics of addressing a long-term issue like climate change.

What is important is that a deadline be set for terminating negotiations on the initial allocation of quotas for meeting the year 2000 objective: a realistic deadline, but one set with reference both to EU processes (notably the 1996 Intergovernmental Conference) and the annual meetings of the Conference of Parties to the UN Climate Convention. This would prevent the process

---

[30] In terms of economic game theory, it transforms the problem from being a one-shot game to being a repeated game. Simulation studies often indicate that the prospects for stable solutions are better in such repeated games.

stumbling on whilst emissions continue to rise, and would clearly establish whether or not Europe has the political will to meet its stabilisation commitment.

6.2 Ensuring and extending a stable system

It is very difficult to predict or control total CO_2 emissions precisely, because of both economic and weather-induced fluctuations in energy demand. With a quota system designed solely to enforce the target of EU emissions stabilisation in 2000 at 1990 levels, what happens if emissions slightly exceed – or fall slightly below – that level?

One approach is to rely on enforcement penalties – 'exceedence charges' – applied retrospectively for the case when emissions exceed quota holdings; and to establish a Community fund which can buy back quotas, perhaps at predetermined prices, if emissions are below the target total. These would in effect define an upper and lower bound to the cost of obtaining quotas. But the approaches would risk unstable fluctuations between these levels, as perceptions varied as to whether emissions would exceed or fall below the total. An attempt to narrow the gap between the lower and upper prices would risk negating one of the benefits of the system, namely that of creating a market system that can itself establish the minimum financial cost of achieving the objective.

An alternative arises again from the observation that the 1990-2000 objective is likely to be but the first step in a much longer process. This opens the possibility of 'banking' quotas for future use. In other words, if under the incentive of the system it proves possible to do better than the stabilisation goal, and the price of quotas drops correspondingly to low levels, parties with spare quotas could elect to 'bank' them for later use – or for selling in the future – based on their expectation of how much emission constraints may tighten after 2000. This both improves the stability of the system, and improves the prospects for exceeding and strengthening the environmental goal. The benefits of allowing such banking has been clearly demonstrated in USA experience with tradeable permits.

This does not necessarily imply that a decision would have to taken initially as to whether the system would extend after 2000; a combination of the two approaches could be considered in which the system could encompass this possibility if, after initial experience, a decision was taken to extend it. But the possibility of a system which extends over time raises a number of other possible design issues. One attractive option could be to establish quotas with extended lifetimes, of which a fraction are retired every few years and replacement quotas issued. The number of replacements quotas, and their distribution, would reflect perceptions of the need to tighten the emissions control and/or expand the range of participants, and the evolving influences upon the ease of constraint in different parties.[31]

6.3 Level and ownership, monitoring and enforcement

A number of definitional issues need to be addressed. First, CO_2 emissions need to be defined on a basis that can be readily monitored. The Commission and OECD guidelines for national reporting of greenhouse gas emissions already establish a definition of CO_2 emissions from commercially traded fossil fuels that can be adequately monitored, assuming there is no deliberate falsification of statistics. The possibility of deliberate fraud – a significant problem in the Community's Common Agricultural Policy – needs to be examined. The fact that fossil fuels are commercially traded and subject to excise taxes suggest that similar institutional systems can be applied to verifying CO_2 emissions, but this is an aspect that should be subject to further study.

An important aspect of designing such a system would be the decision on whether the member governments ultimately 'owned' the quotas, or whether they should in turn be allocated to industries who could trade them freely within the Union.

With government-level quotas, governments would be held accountable to European institutions, by Union law and enforcement procedures, for ensuring that their national emissions did not exceed their quota holdings. The mechanisms by which governments seek to encourage their consumers and industries to limit emissions would remain primarily a matter of national

---

[31] M. Grubb and J. Sebenius, 'Participation, allocation and adaptability in international tradeable permit systems for greenhouse gas control', in *The Economics of Climate Change*, Proceedings of the OECD Conference, OECD, Paris, 1993.

policy, subject to the usual Union competition and related laws and, potentially, Commission initiatives to harmonise specific energy policy/abatement measures, as noted above.

Quota trades would be a matter for negotiation between member states, presumably performed after a great deal of official analysis and discussion. Essentially, this would create a market for CO_2 emissions with the governments as the trading entities. One concern that has been expressed about such a system is the possibility of dominance of the limited market by the largest members. However, by the year 2000, assuming a Union with at least 15 or 16 states, market dominance does not in itself seem likely to pose insuperable problems. Even Germany, by far the largest emitter, would by then account for little over 25% of EU CO_2 emissions; and the next largest, the UK, for under 18%. The remainder would be shared between France, Italy, Spain and a host of smaller countries. It is, however, debateable how efficient a system based upon intergovernmental exchanges would be, though the added flexibility it brings should clearly make it more efficient than having fixed emission targets.

In itself such a system need not involve industry directly. The incentives and institutional mechanisms would apply at the governmental level and governments would act to affect industrial choice with whatever mix of domestic policies they considered appropriate. The implications for industry would be those of knowing with some confidence the EU-wide emissions total, and being able to lobby governments if they consider that certain reductions could be achieved more cheaply by buying quotas from other countries (or vice versa). The price of quota exchanges would begin to give some indications of the value accorded to CO_2 reductions, though a highly imperfect one because quota trades would inevitably have a strong political component.

The alternative would be for governments to allocate (or sell) carbon 'permits' or 'coupons' to their industries, choosing their own criteria for how to share them out; after which governments would have discharged their direct obligations, and the companies could seek to obtain or sell permits anywhere in Europe.[32] The permits would have to be held by major carbon-emitting

---

[32] J. Heister, P. Michaelis and E. Mohr, *The Use of Tradeable Emission Permits for Limiting CO_2 Emissions*, Commission of the European Communities, Brussels, 1992.

industries – big industrial consumers and power generators – and by the energy production and transporting industries themselves for energy delivered to more dispersed applications such as homes, small businesses and transport.[33]

Such a system could be complex, but it would bring the CO_2 trading down to the corporate level and avoid the political complications of intergovernmental quota trading. A carbon permit exchange could be established in Europe much as for other commodity markets, in the same way that the Chicago Board of Trade has begun to act as the exchange for permits to emit sulphur dioxide in the USA. Indeed, in designing such a corporate-level system, some lessons could be drawn from the USA experience with sulphur dioxide control, and proposals elsewhere for CO_2 permit schemes.[34]

The advantage would be that such a system should in principle be more efficient than intergovernmental trading: the industries involved, many of which already operate on a pan-European basis, could make their own judgements about the optimal choice of technologies and location, taking into account the implicit cost of CO_2 emissions revealed by the cost of obtaining permits anywhere in Europe. Furthermore, a system of overlapping permits with extended lifetimes (a possibility noted in the previous section) would allow industres to obtain, at a premium, long-lived permits to reduce the financial risks involved in long-lived carbon-intensive investments.

But there could also be disadvantages to bringing the system down directly to the corporate level. By the same token, it would increase the regulatory complexity faced by industries that already feel burdened by regulations. In its initial stages, such a system could create additional uncertainties about how it would work in practice. The opportunities for fraud could be increased, probably necessitating further regulatory oversight. And such a corporate-level system would reduce the incentive for other aspects of government policy to limit emissions, and could raise other problems such as the protection of

---

[33] For the fossil fuel industries, such a system would probably have some administrative characteristics in common with Value Added Taxation, with corporations able to reclaim carbon credits when fuel is sold to industries – like power generation – that themselves would be covered by the scheme.

[34] For example, Merete Heggelund, *Emissions permit trading: a policy tool to reduce the concentration of greenhouse gases*, Canadian Energy Research Institute, Calgary, January 1991.

new entrants against discriminatory withholding of permits by established companies.

Ultimately, the issue is one of creating the institutional structures required to establish a new market – a market that gives value to avoided CO_2 emissions and hence can counterbalance the current financial interests in increasing fossil fuel consumption. Efficient and effective markets are not created from nothing: they evolve from simpler origins.[35] The extensive and complex legal and informational structures that underpin current commodity markets are the product of many decades of evolution.

For CO_2, the first step may well be to establish a framework in Community law that simultaneously gives agreed national emision targets legal status whilst lifting the current implicit ban on trading them between member countries. By providing visible financial incentives, this in itself should ensure that governments take their commitments seriously enough to achieve the Union's declared commitments. Whether, when, and how more complex systems evolve are questions that need not yet be resolved. What matters is to break out of the current impasse and start moving in the right direction.

---

[35] A discussion of this with reference to an international tradeable CO_2 quota system is given by Richard L. Sandor, 'Implementation issues: market architecture and the tradeable instrument', in *Combating global warming*, United Nations Conference on Trade and Development, UNCTAD/RDP/DFP/1, Geneva/New York, 1992.